Stegosaurus

Lori Dittmer

CREATIVE EDUCATION
CREATIVE PAPERBACKS

seedlings

Published by Creative Education and Creative Paperbacks
P.O. Box 227, Mankato, Minnesota 56002
Creative Education and Creative Paperbacks
are imprints of The Creative Company
www.thecreativecompany.us

Design by Ellen Huber
Production by Rachel Klimpel and Ciara Beitlich
Art direction by Rita Marshall

Photographs by Alamy (Panther Media GmbH, Science Photo Library), Bridgeman (Look and Learn), DepositPhotos (Pixelchaos), iStock (jondpatton), Science Source (Science Photo Library, Millard H. Sharp/ Science Source), Shutterstock (Catmando, Daniel Eskridge, Elenarts, Herschel Hoffmeyer, Michael Rosskothen, SciePro, Vac1, Warpaint)

Library of Congress Cataloging-in-Publication Data
Names: Dittmer, Lori, author.
Title: Stegosaurus / by Lori Dittmer.
Description: Mankato, Minnesota : Creative Education and Creative Paperbacks, [2024] | Series: Seedlings: dinosaurs | Includes bibliographical references and index. | Audience: Ages 4–7 | Audience: Grades K–1 | Summary: "Early readers are introduced to Stegosaurus, a Jurassic plant eater with rows of plates. Friendly text and dynamic photos share the dinosaur's looks, behaviors, and diet, based on scientific research"— Provided by publisher.
Identifiers: LCCN 2022015663 (print) | LCCN 2022015664 (ebook) | ISBN 9781640265042 (library binding) | ISBN 9781682770566 (paperback) | ISBN 9781640006348 (ebook)
Subjects: LCSH: Stegosaurus—Juvenile literature. | Dinosaurs—Juvenile literature.
Classification: LCC QE862.O65 D585 2024 (print) | LCC QE862.O65 (ebook) | DDC 567.9153—dc23/eng/20221026
LC record available at https://lccn.loc.gov/2022015663
LC ebook record available at https://lccn.loc.gov/2022015664

TABLE OF CONTENTS

Hello,
Stegosaurus!

This dinosaur lived long ago.

Diplodocus and *Allosaurus* lived at the same time.

Stegosaurus fossils were first found in 1876. It seemed like its bony plates lay flat.

But they didn't.

They stood straight up from its back.

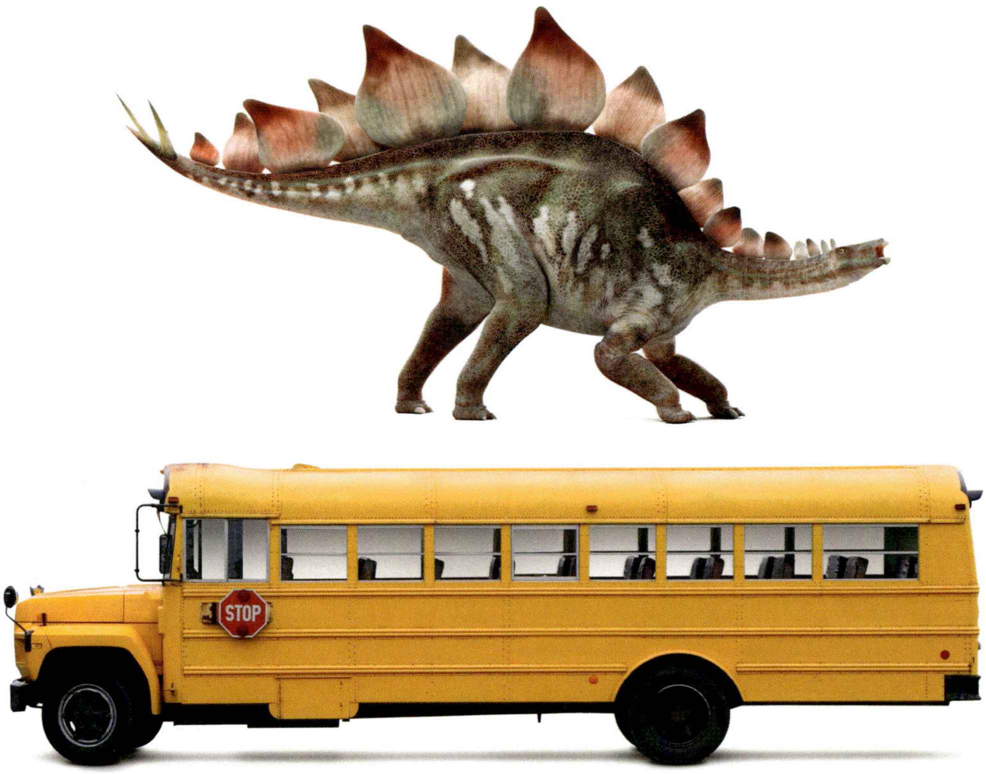

Slow *Stegosaurus* walked on four legs. It was almost as long as a bus.

But it had a small head and a tiny brain.

Long **spikes** stuck out at the end of the tail.

Stegosaurus may have used the spikes to fight enemies.

Stegosaurus ate plants using its beak.

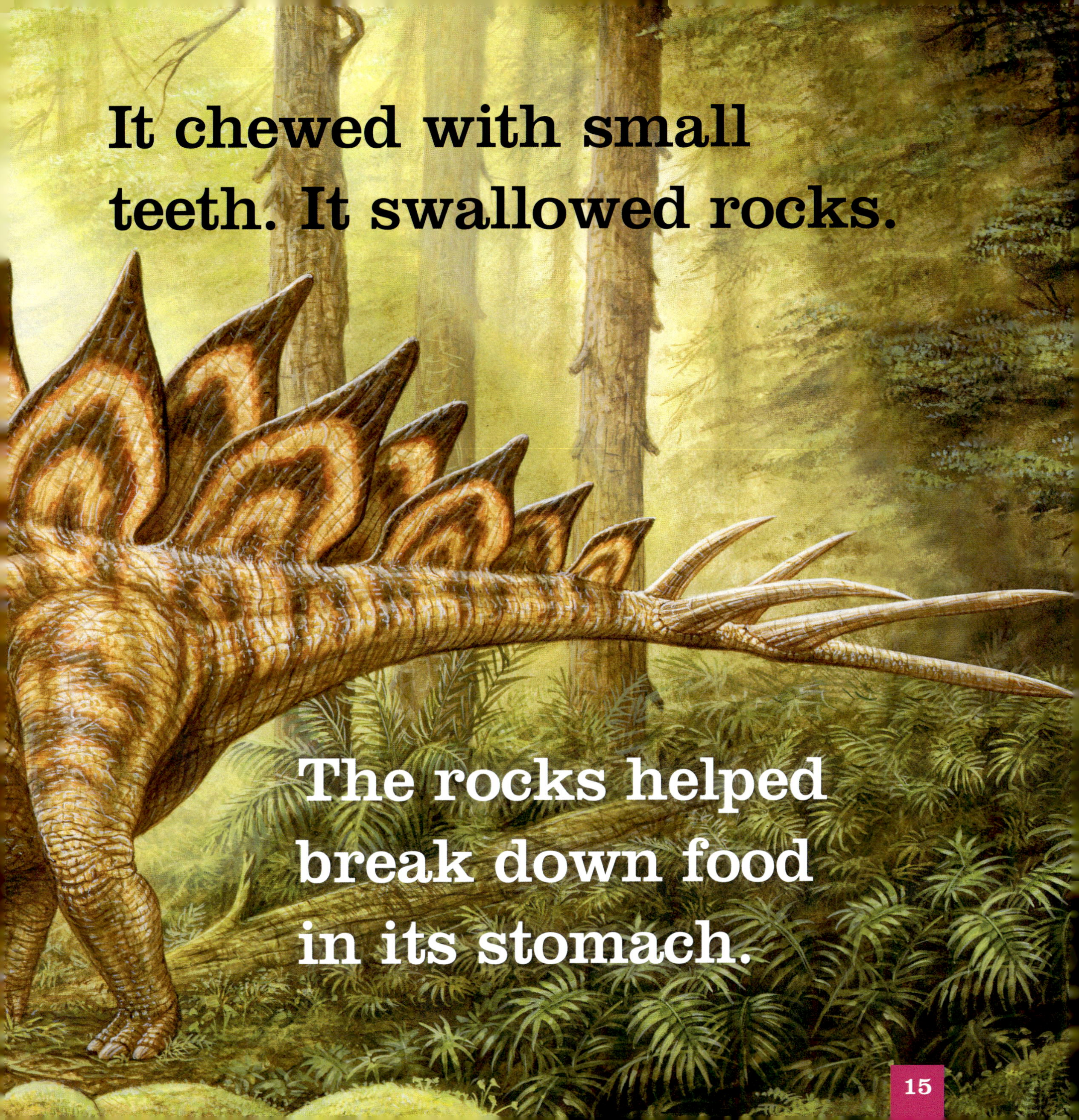

It chewed with small teeth. It swallowed rocks.

The rocks helped break down food in its stomach.

Stegosaurus held its head low.

It found food near the ground.

Goodbye, *Stegosaurus*!

Picture a *Stegosaurus*

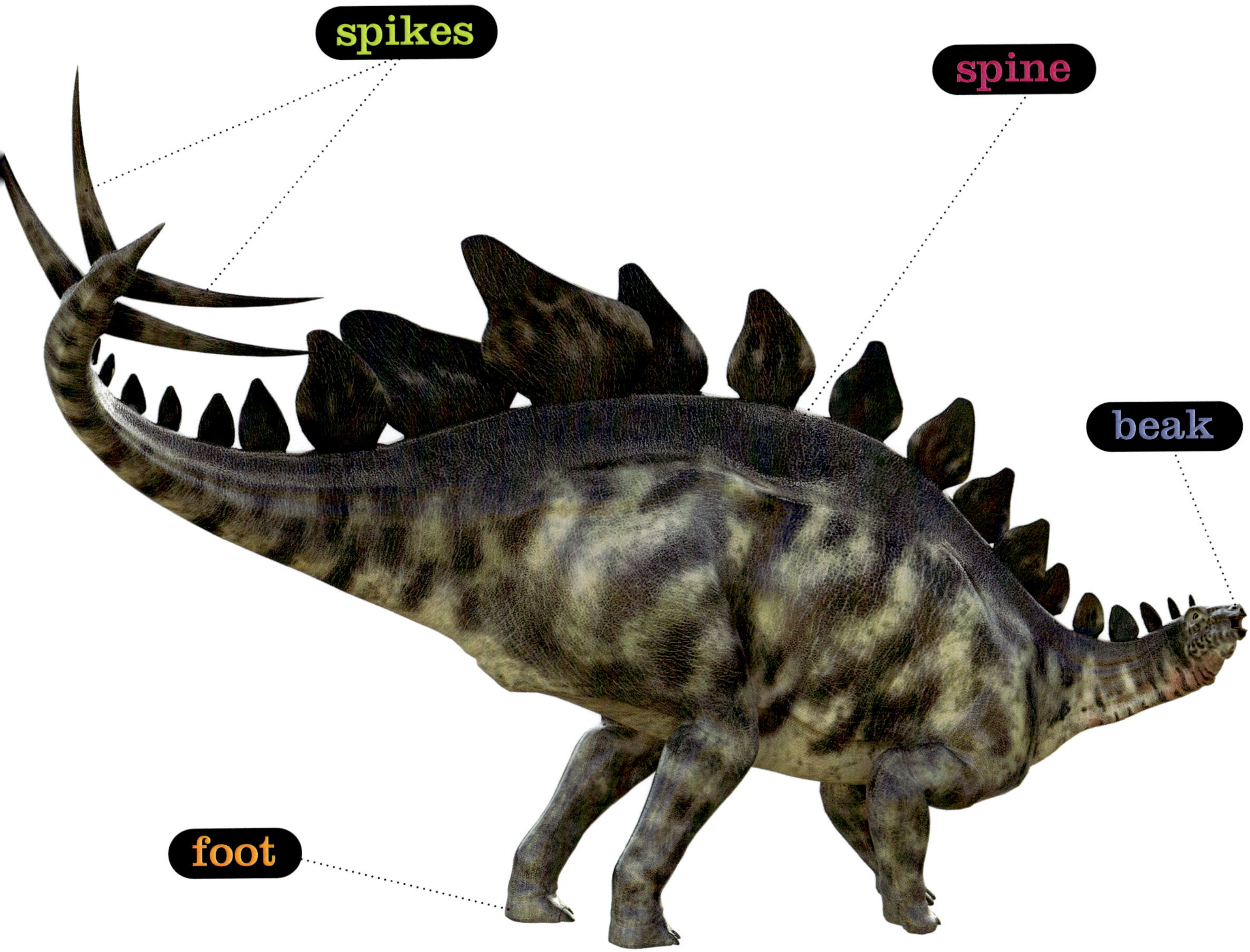
spikes
spine
beak
foot

Words to Know

beak: the hard, pointed part of the mouth of some animals

fossil: a bone or trace from an animal long ago that can be found in some rocks

plate: a flat bone

spike: a sharp, pointed object

Read More

Colins, Luke. *Stegosaurus*. Mankato, Minn.: Black Rabbit Books, 2021.

Golusky, Jackie. *The Mighty Stegosaurus*. Minneapolis: Lerner Publications, 2022.

Websites

DK Find Out! | *Stegosaurus*
https://www.dkfindout.com/us/dinosaurs-and-prehistoric-life/dinosaurs/stegosaurus
Read more about *Stegosaurus* and take a dinosaur quiz.

PBS Eons | Stegosaurs: Tiny Brains & Thagomizers
https://video.idahoptv.org/video/stegosaurs-tiny-brains-thagomizers-3ketlj
Watch a video about *Stegosaurus* and its relatives.

Note: Every effort has been made to ensure that the websites listed above are suitable for children, that they have educational value, and that they contain no inappropriate material. However, because of the nature of the Internet, it is impossible to guarantee that these sites will remain active indefinitely or that their contents will not be altered.

Index